Mobil New Zealand Nature Series

MUSHROOMS AND TOADSTOOLS

Mobil New Zealand Nature Series

MUSHROOMS AND TOADSTOOLS

Marie Taylor

REED

First published 1981
A. H. & A. W. Reed Ltd
68–74 Kingsford-Smith Street, Wellington 3
53 Myoora Road, Terrey Hills, Sydney 2084
also
16–18 Beresford Street, Auckland
85 Thackeray Street, Christchurch 2

ISBN 0 589 01354 8

Typesetting by Printset Processes (1973) Ltd., Christchurch

Contents

Introduction

Most of us have had the experience of coming upon a cluster of fungi in a place where we have never seen them growing before. They arouse our curiosity. We may keep returning to the spot but we may never see that particular species in that place again. Why do they spring up so briefly only to disappear? Are they harmless or dangerous? It seems that fungi are one of the least understood groups of organisms, even among people with a particular interest in natural history. One explanation is that fungi are ephemeral. This book is designed to answer questions about the life and habits of these fascinating but neglected plants.

PLANTS OR NOT PLANTS?

It is traditional to regard fungi as plants when common sense tells us they are not animals. They grow wherever plants are found, and they are included in the botanists' domain of inquiry, but they are not part of the green cover of the earth. Among micro-organisms the distinction between plants and animals becomes blurred. In the case of fungi, they are sedentary organisms with a branching, filamentous growth form. They reproduce and are dispersed mainly by spores. In these characters they resemble simple plants. On the other hand their eating habits are decidedly unlike plants'. Plants are producers, feeding themselves by photosynthesis, getting their energy from sunlight and their carbon from the carbon dioxide of the air. Fungi are consumers depending on food and energy from complex organic compounds derived from plants, animals, or their products. They break down organic matter into energy and soluble foodstuffs by means of enzymes. The principles are the same in animal nutrition, but animals take their food inside their bodies for digestion, whereas the frailer fungus digests the food outside itself and absorbs only the particular soluble products it requires. Fungi have a powerful ability to break down large molecules and alter the chemistry of the substrate they are growing in, a process we call decay.

Fungi probably diverged from the first primitive living cells along a separate line from either plants or animals; and the best way to classify them is to place them in a fungus kingdom with a separate status of its own. Nevertheless, it is convenient to go on referring to them as plants.

WHAT IS A FUNGUS?

The structure that we commonly call a fungus is the spore-bearing fruitbody of an organism which spends the greater part of its life buried from sight. The true body of the fungus is a microscopic jungle of cobwebby threads called hyphae and the aggregate tangle of hyphae is the mycelium. The mycelium is occasionally matted thickly enough to be seen with the naked eye, as under the bark of a rotting log, or in mushroom compost, when it is known as spawn.

THE LIFE CYCLE OF A FUNGUS

Before a fungus can complete its life cycle it is necessary for two compatible strains of hyphae to meet and mingle together into a single mycelium. This mating produces secondary mycelium, capable of forming fruitbodies and finally, spores. It is the fruitbodies in their various forms of mushroom, puffball, earthstar, bracket ("conk"), morel or pixie cup which are pictured in this book.

Do not forget that these fruitbodies are only a launching pad for spores. They are the way the mycelium reproduces, the culmination of the unseen growth, feeding and mating functions of the mycelium. Many fruitbodies have no more than a few days of life; tougher kinds may last a week or two. Some woody bracket fungi are perennial and continue releasing spores over a period of years. A glance at the illustrations will show that some fruitbodies are green, but the green colour is never due to the presence of chlorophyll.

Spores are tiny, single cells, no bigger than the specks of dust in a sunbeam. They can travel long distances and are found in air samples taken high up in the earth's atmosphere. They are produced in astronomical numbers to offset the slim chances they have of landing in a suitable unoccupied habitat. Even when a spore does germinate and become established as a new mycelium, there is still the hazard of finding the right mating partner before the life cycle can be completed. This is one reason why so many people are disappointed in their attempts to naturalise mushrooms in their lawn. They hopefully scatter pieces of mushroom here and there, when they might have a better chance of success if they buried some fruitbodies together in a richly composted soil.

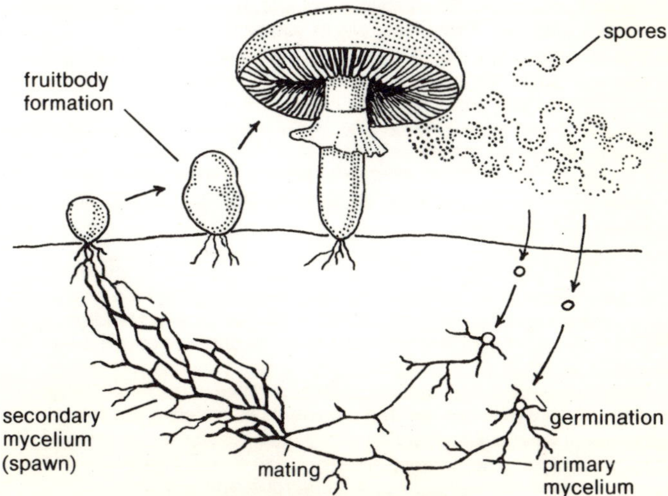

LIFE CYCLE OF A MUSHROOM

TO EAT OR NOT TO EAT?

Many fungi are of no gastronomic interest at all because they are too small or too tough, or because they have an unpleasant taste. But leaving these aside there still remain a number of fleshy species that are edible and good eating. Poisonous species are a minority, but must be taken into account. It is wise to begin by learning to recognise the harmful ones, especially *Amanita*, and avoid them when collecting wild mushrooms for the table.

Numerous pieces of folk wisdom claim to give reliable tests for the recognition of poisonous species, or suggest treatments like salting or pre-boiling which guarantee to remove poisons. All of these claims are false. There is no simple rule for distinguishing edible from poisonous, and no treatment will render the deadly species harmless. One must learn the structure and appearance of each species to be certain of its identity.

On the other hand it is not necessary to go to extremes. No harm can come to you through handling a fungus, and there is little point in kicking fungus fruitbodies to pieces in the mistaken belief that their very existence is a hazard.

Recalling the fungus life cycle you will see that you cannot damage the parent mycelium by destroying fruitbodies. In fact, if the spores are ripe you only help to disperse them.

COLLECTING FOR THE TABLE

1. Be sure of your identification. Check to see if there are any odd men out and throw away doubtful ones.
2. Choose only young, fresh specimens. Old, weathered specimens may contain maggots. Trim off the dirt as you pick.
3. Use an open container to keep your mushrooms in good condition until you get home. Never use plastic bags in which the mushrooms will "sweat" and possibly spoil.
4. Eat with moderation the first time, in case you have a personal allergy to a good edible species.
5. Field mushrooms do not need to be peeled. A wipe with a damp cloth should be sufficient preparation. Boletes should be peeled and the tough stalk removed.

In New Zealand we eat the cultivated mushroom, the field mushroom (pink button) and the horse mushroom. Chinese New Zealanders enjoy the ear fungus. Many other kinds are eaten in different countries of the world.

Almost nothing is known of the edible qualities of the native fungi from the New Zealand bush. The comments in this book are made on the basis of what is known about the species we share in common with other countries. It cannot be emphasised too strongly that eating any fungus you are not familiar with is dangerous, and it is your responsibility to decide whether the risk is worth it.

FUNGAL ASSOCIATIONS WITH INSECTS

Sooty moulds are common in bush, scrub and gardens. Some look like a coating of black dust on the leaves, others take the form of whiskers or hairy black cushions on the twigs. They grow on the host as indirect parasites, taking nourishment from the surplus honeydew exuded from aphids and scale insects which suck the plant's sap, and they may harm the plant by shading its leaves from the sunshine they need for photosynthesis. The presence of the mould is indicated by a delightfully sweet smell in the vicinity. Sooty moulds often grow on garden varieties of manuka, and the best cure is to spray with insecticide to remove the insects supplying the mould with food.

Septobasidium is a jelly fungus which grows as a waxy white crust surrounding the twigs of lancewood and five finger (*Pseudopanax*). Under the crust roof are spaces occupied by scale insects, some of which are normal and healthy, while many others are prisoners held in shackles of parasitic fungal hyphae, unable to move or to reproduce. The fungus is siphoning away the sugary juice they extract from the plant and diverting it for its own nourishment.

"Vegetable wasps" and "plant worms" have given rise to tall stories about insects growing into trees, when in fact they are simply insects attacked by fungal parasites. *Cordyceps* (Plate 31) and the vegetable cicada fungus *Isaria sinclairii* (Berk.) Petch are examples. The underground root-feeding cicada nymph is penetrated by hyphae of *Isaria* which live on its tissues and kill it, later sending up a branched white stalk to produce spores above the soil surface. The clusters of fluffy white spores (conidia) look like a pinch of flour spilt on the leaf mould.

Sooty mould fungus on a twig

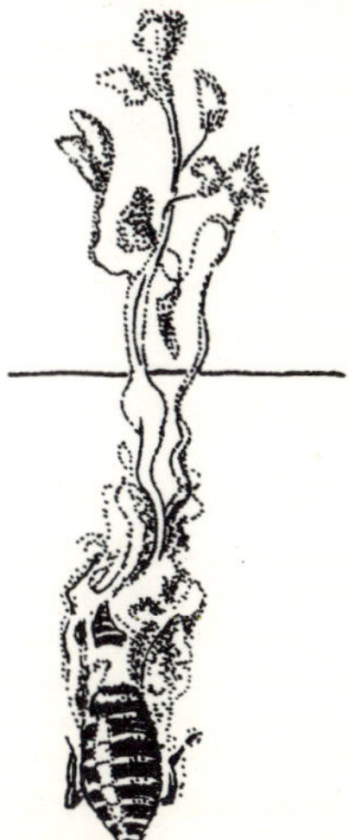

Vegetable cicada

In autumn one can find dead bush and garden insects which have fallen prey to yet another parasitic fungus. These victims have white patches or bands of fungus erupting through the softer chinks between their body plates, giving them the appearance of being frosted with icing. The disease is known as sugar bonbon disease, and it attacks adult cicadas, mantids, stick insects, vegetable bugs, wasps and sometimes spiders.

FUNGI IN NEW ZEALAND

Some of our fungi are members of a group having a worldwide distribution and others are known to have been introduced with plants or soil from overseas. These are the common fungi of gardens, parks and farmland, and they can often be identified from European and American books. Besides these, there are numerous indigenous species found in association with the native vegetation. The New Zealand fungus flora is not completely known and therefore quite a number of species are still without scientific names.

At the present state of knowledge and within the confines of a field guide using no microscopic or chemical characters, it is possible to make only a selection of common and conspicuous species. Additional information given in the text opposite each plate should help you to place your finds in the right group. Don't be put off by the Latin names; after all, you use them for *Dahlia, Fuchsia* and *Chrysanthemum*. It is better to tackle the scientific names than to invent false names which are only a translation of the Latin one.

Fungus hunting with camera or sketchbook can be a great sport. It takes you out walking in the open air — as far as you like at whatever pace you like. There is a spice of uncertainty about the hunt. There is always the possibility of making a new find in an old place, and like the fisherman you will sometimes come home empty handed. In this hobby it is not even necessary to go far afield: one's own garden can spring surprises.

Do remember the nature code and take no more than one or two fruitbodies to show different stages of development. While it is true that fungus collecting does very minimal harm to the environment, discriminating people are aware that the biological world is in a fragile balance and they remove only what they intend to make use of.

THE ROLE OF FUNGI IN NATURE

It is not always appreciated that the lowly fungi have an essential place in the economy of nature. They, together with the bacteria, are agents of decay. We humans wrinkle up our noses at the thought of organisms which get their living by breaking down the bodies of plants and animals, and by spoiling organic products, particularly products useful to us. But the decomposers ensure that complex compounds are returned to the soil and air in mineral form, so that the basic atoms of life can be taken up and used once again by new generations of plants and animals. Without this natural recycling process in action, life on earth would come to an end.

A Picture Guide to the Classification of Fungi

A. ASCOMYCETES produce their spores inside a fertile hypha called an ascus (see drawing). Each ascus commonly produces eight ascospores inside it, and when ripe they are squirted out through the ascus tip into the air. They are flung clear of the fruitbody to where the slightest turbulence will carry them away. Ascomycetes are more numerous than basidiomycetes but few of them produce conspicuous fruitbodies.
Plates 31 and 32

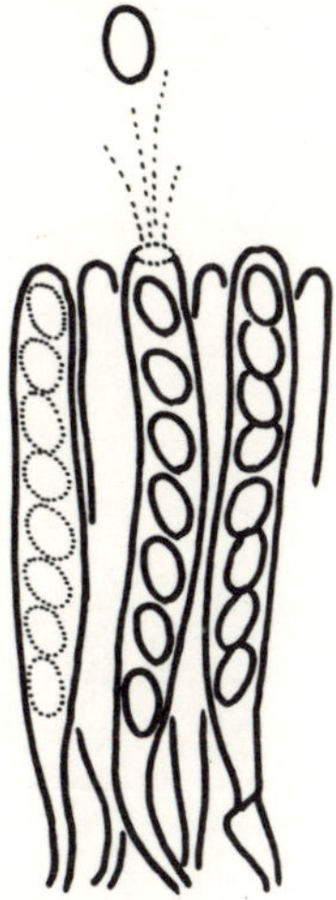

Microscopic detail of asci much enlarged.

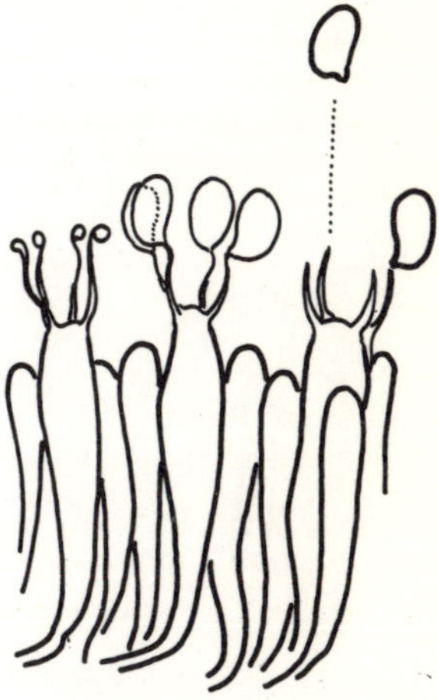

Microscopic detail of basidia much enlarged.

B. BASIDIOMYCETES form spores on basidia (see drawing). The basidium produces four basidiospores borne on the tips of little prongs which, projecting from its apex, lift the spores clear of the surrounding surface.
Plates 1–30

B1. HYMENOMYCETES have an exposed sporing surface from which each basidiospore in turn is actively shot away.
Plates 1–22, 26–30

B2. GASTEROMYCETES have the basidia enclosed during development. At maturity, either the fruitbody remains closed, or if it opens it releases spores by a method different from the one in Hymenomycetes.
Plates 20, 23–26.

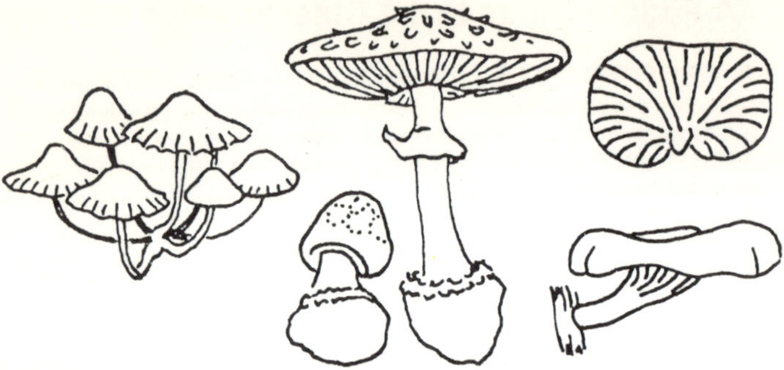

Agarics Plate 1–19, 21 Fleshy with gills; with or without stalk

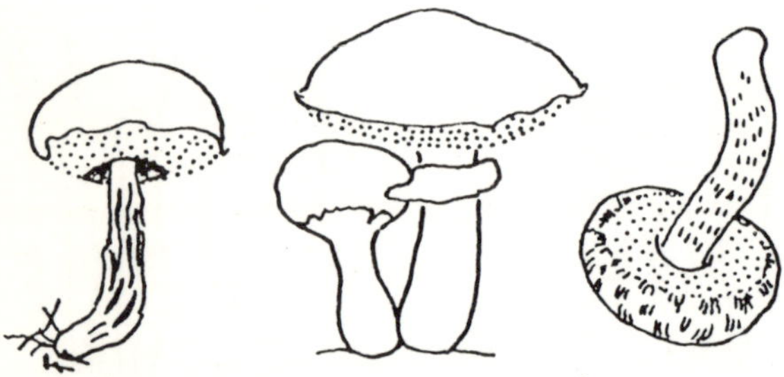

Boletes Plate 21, 22 Fleshy with pores; stalk present

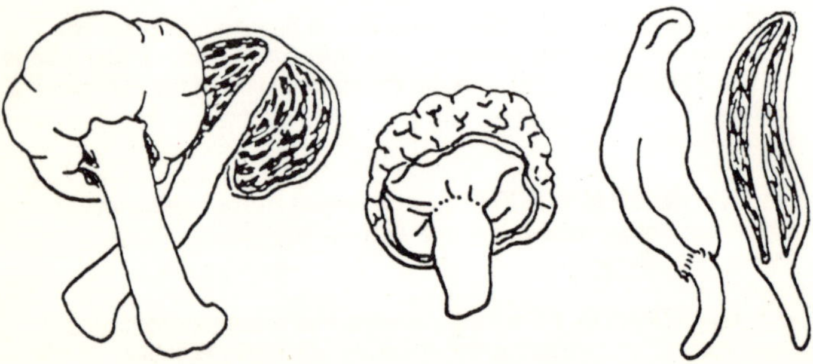

Secotiums, Pouch Fungi Plate 20 Fleshy; stalk inserted into closed cap; spores not actively dispersed

Puffballs, Earthballs, Tuberous Fungi Plate 23, 24 (also 20) Round or oval spore sacs, base sometimes sterile
Earthstars Plate 24 Star shaped with puffball in centre

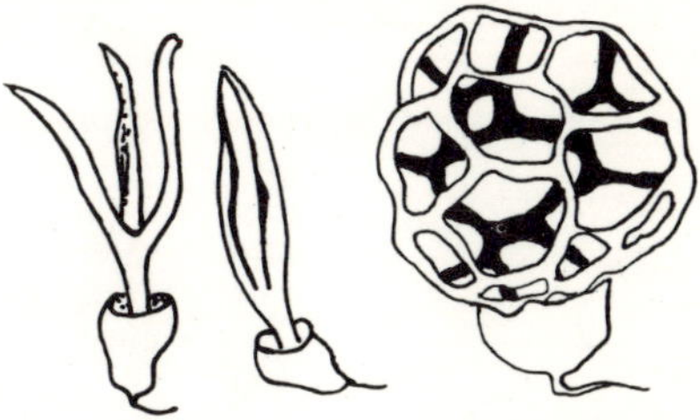

Stinkhorns Plate 25 Opening to display olive-brown slimy spores; smell unpleasant

Birdsnests Plate 26 Like a nest containing eggs

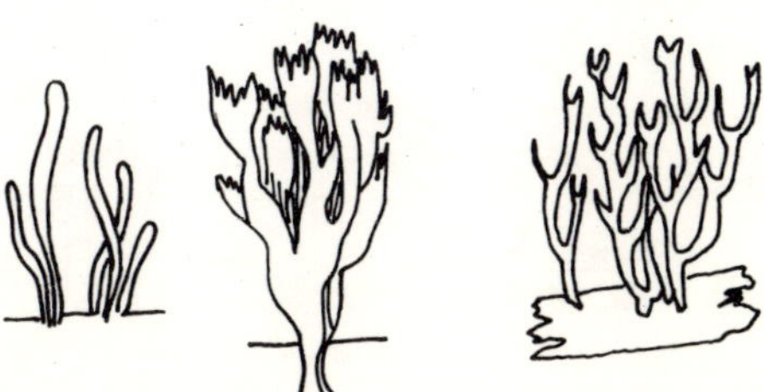

Coral Fungi, Fairy Clubs Plate 26 Like coral, single or with many erect branches

Polypores Plate 27 Tough to woody, with pores; with or without stalk, growing on wood

Spine or Tooth Fungi Plate 28, 30 Fleshy to tough, with hanging spines or teeth

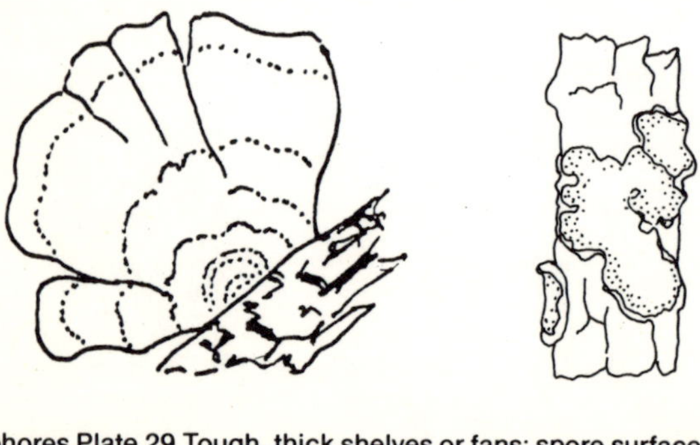

Thelephores Plate 29 Tough, thick shelves or fans; spore surface smooth or slightly irregular
Paint Splash or Crust Fungi Plate 29 Tough, thin crusts; lying flat, forming shelves or fan shaped; spore surface smooth; on wood

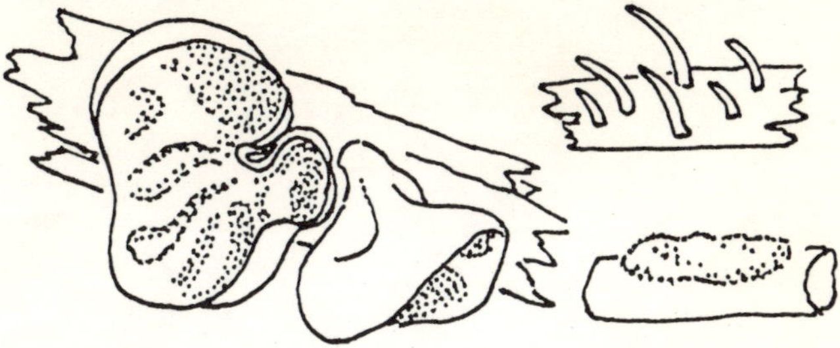

Jelly Fungi Plate 30 Gelatinous or rubbery; on wood

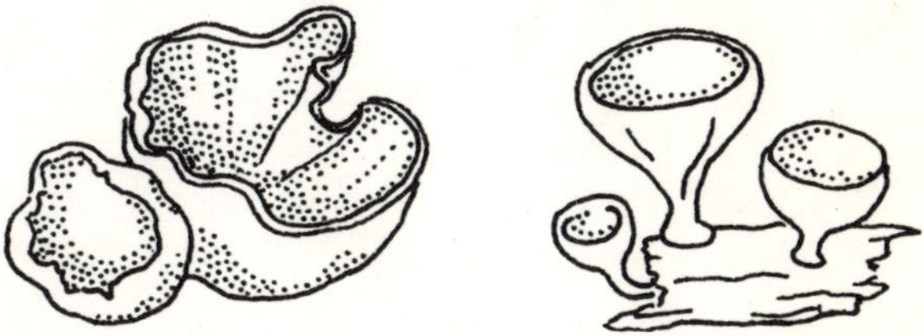

Cup Fungi Plate 31 Cup shaped, bowl shaped or flattened discs

True and False Morels Plate 32 Pitted or wrinkled head with stalk, fleshy

1

2

The deadly Amanitas

Amanita is the genus to which the most dangerously poisonous species belong, and one should begin by learning to recognise and avoid them before gathering any wild mushrooms for food. The important characters to notice are the white spore powder (spore print, see p. 27), gills free from the stem, and the presence of a universal veil. In some species a partial veil is present too.

The universal veil is an outer sheath completely surrounding the immature button toadstool. During expansion it is ruptured and its remnants are left on the base of the stalk and on the cap. If the veil tissue is tough and elastic it breaks cleanly, leaving a torn membranous cup (the volva) around the base. (See *A. phalloides*.) In *A. muscaria* the universal veil tissue is fragile and crumbles to pieces under strain, the upper part forming the white patches on the cap, and the lower part forming irregular bracelets of warts on the basal bulb.

Both of these *Amanita* species have an inner membrane which covers the young gills. When this partial veil is pulled away from the cap margin it collapses around the stalk like a skirt. In the mature fruitbody the partial veil becomes the ring.

1. *Amanita phalloides* (Vaill. ex Fr.) Secr. Death Cap. *Deadly poisonous*. One teaspoonful of this fungus is lethal. It is found growing near oak trees, and usually has a greenish or olive-brown cap. Sometimes the cap is whitish but it cannot be confused with the edible field mushroom if the collector takes care to dig up the fungi and examines the foot of each for signs of the telltale volva. When overmature the Death Cap has a sweetish, nauseating smell.

2. *Amanita muscaria* (L. ex Fr.) Hooker. Fly Agaric. *Poisonous*. Like the Death Cap this species is an introduced one in New Zealand, and it associates with pines, oak, silver birch or sweet chestnut. By crushing the cap in milk or sprinkling it with sugar one can make a poisoned bait for killing flies, hence the name Fly Agaric.

The appearance and colour of this toadstool are somewhat variable, as rain may wash out some of the red pigment or remove the white veil patches.

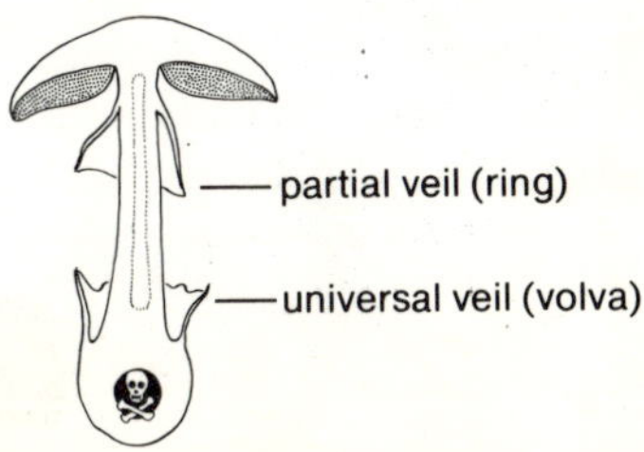

Amanita phalloides — diagram of structure

3
4
5
6
7

Amanita species native to New Zealand

These are *Amanita* species found in association with New Zealand plants. Note carefully the different appearances taken by the universal veil, from a membranous sac to a jutting rim dusted with dark, powdery fragments. In the absence of evidence to the contrary, regard all Amanitas as poisonous.

3. *Amanita* sp. *Poisonous*. This species associates with *Leptospermum*.

4. *Amanita* sp. Yellow, warty *Amanita*. *Poisonous*. This species associates with *Leptospermum*.

5. *Amanita* sp. *Poisonous*. A ringless species, with a dark line along the gill margins. This species associates with *Nothofagus*.

6. *Amanita nothofagi* Stevenson. *Poisonous*. This is a large species with cap up to 11 cm in diameter. It associates with *Nothofagus* or *Leptospermum*.

7. *Amanita farinosa* Schw. *Poisonous*. This powdery grey *Amanita* is found also in western North America, Japan and China. It associates with *Leptospermum* here.

FUNGUS ROOTS

It is thought that most, if not all, forest trees and many shrubs have an association between their roots and fungal hyphae, and several kinds of mycorrhizas or fungus roots are known. In conifers, oak and beeches, the hyphae weave a sheath around the rootlet and penetrate between its outer cells. They are of vital importance to the tree since what it absorbs from the soil passes via the fungus. The fungal hyphae act as extensions of the root into the soil, and the enzyme activities of the fungus enhance the root's ability to obtain mineral nutrients like nitrate and phosphate. In turn the fungus gets a supply of sugars from the tree. Mycorrhizas are of special interest in forestry because they are critically important to the growth of trees on poor soils. Research is directed towards finding the most beneficial strains of fungus so that seedling trees can be inoculated with them before planting out.

Nothofagus and *Leptospermum* of the New Zealand flora have as mycorrhizal associates species of *Amanita*, *Russula*, *Cortinarius* and *Tylopilus*. In autumn the most spectacular fruitings of fungi in this country are in beech forest and *Leptospermum* shrubland.

8
9
10

Lepiota and *Macrolepiota* — the parasol mushrooms

Parasol mushrooms have white gills, free from the stem, and a ring placed half way or higher on the stem. They have white spores which redden in iodine solution (dextrinoid reaction). The cap has an upper surface layer, often coloured differently from the flesh, and during expansion this skin is torn into scales surrounding a central, intact disc. Note that some features resemble those of *Amanita*, but *Lepiota* has no universal veil.

There are scores of small, fragile *Lepiota* species in which the ring is easily lost. They grow as saprophytes (decomposers) in all kinds of forest. The small species have no interest as food, and some of them are reported to be *poisonous*.

8. *Lepiota* sp. Regard as *poisonous*.

9. *Macrolepiota* sp. Parasol Mushroom. *Edible*. The cap is up to 10 cm across, set on a tall, slim stalk. It is easy to recognise, although you must make sure that the spore print is white (see p.27). There is a poisonous genus *Chlorophyllum* of similar general aspect but with a pale green spore print.

10. *Macrolepiota rhacodes* (Vitt.) Singer. Shaggy Parasol Mushroom. *Edible*. This mushroom can attain 15 cm across, sometimes more, and the cap scales are much more deeply cut than in the Slim Parasol. The mushroom has thick, white flesh which turns pinkish-orange on cutting. Other features are the double ring with fringed margins, and the removable stem which parts easily from the cap leaving a circular depression in the cap flesh. The cap colour is variable, from pale fawn through red-brown to deep brown in different fruitbodies. This species is found in rich humus in gardens and parks or under trees, especially pines, but it is not known to be mycorrhizal.

11

12

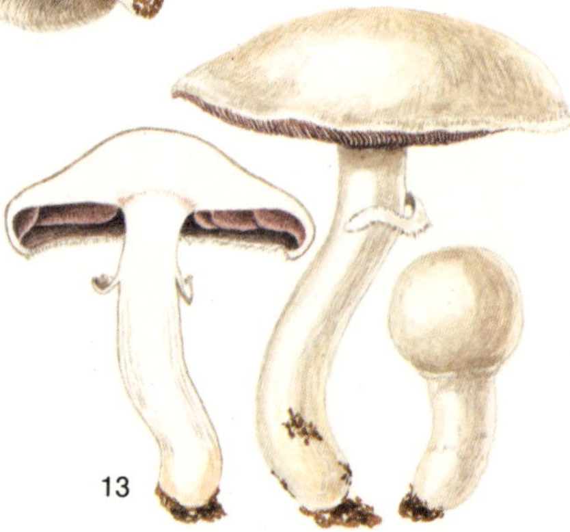
13

14

Agaricus — the meadow mushrooms

Many species of *Agaricus* are well known, edible mushrooms. One small group of the genus cannot be recommended for eating because they cause stomach upsets, but none are fatal. (See Plate 5 for these.)

Agaricus is recognised by free gills that are pink in the young button and gradually darken to purple-brown or chocolate as the spores mature. The stalk always has a ring, usually persistent and easily seen, and the spore print is purple-brown (see p.27).

The cultivated mushroom, *Agaricus bisporus*, is derived from a wild mushroom of Europe, possibly *Agaricus hortensis*. After so many years of cultivation it has evolved into a distinct species, with a more delicate flavour than the wild kinds.

11. *Agaricus bitorquis* (Quél.) Sacc. Two Ringed Mushroom, Little Pink Mushroom. Pale form. *Edible*. This is the common small mushroom of grassy places. It has a double partial veil and the stalk has some cottony fluff on the lower part below the two ring traces. The general shape is squat, with a thick cap, pinkening flesh and pointed foot.

12. *Agaricus bitorquis*. Dark form. *Edible*. Sometimes the cap bears fawn or dark brown surface fibrils instead of being smooth and downy white.

13. *Agaricus campestris* L. ex Fr. Field Mushroom. *Edible*. This is not as common as the little pink mushroom, from which it differs in having a single membranous ring.

14. *Agaricus arvensis* Schaeff. ex Secr. Horse Mushroom, Snowball Mushroom. *Edible*. This is a large mushroom which can be up to 18 cm across and its gills are pale fawn-grey when young, not pink. The cap is white at first and ages to creamy beige with yellow tints where touched. The ring is double, and consists of a smooth upper membrane with a circle of thick, cottony patches marking the tips of a cogwheel pattern on its lower surface. The scent is like aniseed. Because of the yellowing and the chocolate-brown gills some people mistakenly think that this mushroom is not edible.

15

16

17

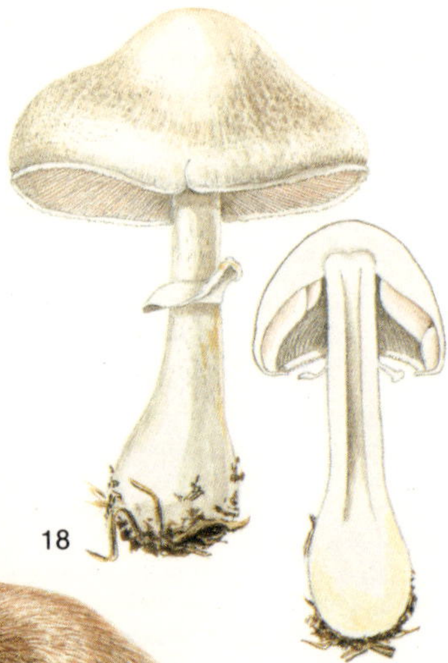

18

19

More *Agaricus* and *Leucoagaricus*

15. *Agaricus* sp. aff.* *meleagris* Schaeff. *Not recommended*. Flesh yellowing at base of stalk on cutting. Cap flesh very thin.

16. *Agaricus* sp. aff. *xanthodermus* Genevier. *Not recommended*. Cap flesh very thin. Not yellowing as strongly as true *A. xanthodermus*. Strange smell, like ink or phenol.
Both the above grow in lawns and gardens, near trees.

17. *Agaricus* sp. Grey Speckled Bush Mushroom. *Not recommended*.

18. *Leucoagaricus naucinus* (Fr.) Singer. *Edible but not recommended*, as it could be confused with *Amanita*. Cap wholly white or with fine traces of grey fibrils in age. Gills white, becoming pink and finally grey when old. Spore print white.

19. *Agaricus subperonatus* (Lange) Singer. *Edible*. A large, fleshy species with cottony scales on the stalk below the ring.

FAIRY RINGS

The circles of mushrooms and puffballs known as fairy rings are seen most readily in grassland but they are also found under forest. The fruitbodies are arranged in a ring because they have sprung from a mycelium growing radially outwards through the soil. The mycelium advances year by year into fresh territory, while the part left behind dies. Fruitbodies are produced just inside the growing rim where the mycelium is best fed. The rings of some mushroom species are accompanied by a zone of taller or greener grass. In undisturbed soil, fairy rings may persist for a long time and the size of some measured in Colorado showed they must have been growing there for about 400 years.

The fairy ring mushroom, *Marasmius oreades*, kills the grass and creates a bare zone where its mycelium is active. It can be an unsightly pest in lawns and playing fields. The fruitbodies of this mushroom are pale cream to beige in colour with a tough, leathery texture and twisted stalk, the cap is bell-shaped and the spore print is white.

* aff. = closely resembles.

20

21

22

23

24

Agaricus species native to New Zealand

All of the species on this page were found in native vegetation and their edibility is *unknown*.

20. *Agaricus variegans* Moll. A mushroom with pinkening flesh in the cap and upper stalk.

21. *Agaricus* sp. This species has the ring set low on the stalk.

22. *Agaricus* sp. Cap flat at maturity with zones of radial black fibrils.

23. *Agaricus* sp. aff. *purpureoniger* Heinemann. A member of a special group of *Agaricus* with the stem woolly or furry below the ring.

24. *Agaricus bambusae* var. *australis* Heinemann.

MUSHROOM GROWTH

The growth of fungus fruitbody from mycelium (spawn) proceeds slowly from the first aggregation of hyphae until the button is formed. Inside the button are all the structures of the adult fruitbody in miniature, and when favourably warm and moist conditions occur the button takes up water and swells to full size in a matter of hours. Rapid expansion of tissues already formed in the button accounts for the dramatically sudden appearance of many fungi, which we term "mushroom growth". Expanding fruitbodies can exert considerable pressure, and many examples are known where paving slabs have been pushed up by apparently soft and fragile fungi.

MAKING A SPORE PRINT

To determine the spore colour of an agaric, cut off the stalk just below the gills and place the cap, gills down, on a piece of clear plastic film. Paper can be used but white spores may not show up very well on it. Transparent film can be viewed against different backgrounds to determine whether the spores are truly white or are a pale tint of pink, cream or green. Cover the cap with a glass or place in a closed container to keep it moist. In a few hours or overnight the spores will collect in a pattern corresponding to the gill spaces. The print may be kept as a record of spore colour.

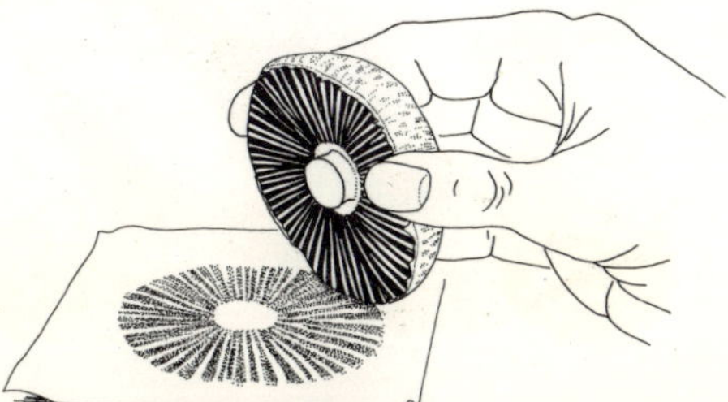

25
26
27

Armillaria — the honey mushrooms

In *Armillaria* there are a number of species which are parasites, attacking and killing living trees. Once established, the mycelium spreads throughout the trunk and roots of the host tree, and when the tree dies the fungus continues to flourish and produce fruitbodies from the dead wood as a saprophyte. *Armillaria* forms a network of tough black strands rather like boot-laces growing between the bark and the wood. These rhizomorphs are unique and can be used to diagnose the presence of *Armillaria* whether the mushrooms are present or not. They are a powerful factor in the success of Armillarias as parasites, as they conduct food and water and enable the fungus to spread through the soil using a former host as the food base to support an attack on the roots of neighbouring trees. The name honey mushroom refers to the colour of the fruitbodies.

25. *Armillaria novae-zelandiae* Stevenson. This species has a moist to slippery cap when fresh. It is dark when young, and as the cap expands the outer skin is broken up into a scatter of dark granules. In age, the cap and gills become blotched with pinkish brown. There is a tenuous white veil which may be seen as a ring, but it soon withers and disappears.

26. *Armillaria* sp. This has a yellow cap with a powdery dry surface.

27. *Armillaria limonea* Stevenson. A large species with a dry cap that feels like sandpaper. It is tough and robust, and the stem below the substantial membranous ring is velvety or sometimes bristly. The yellow cap fades to cream or whitish with age.

28

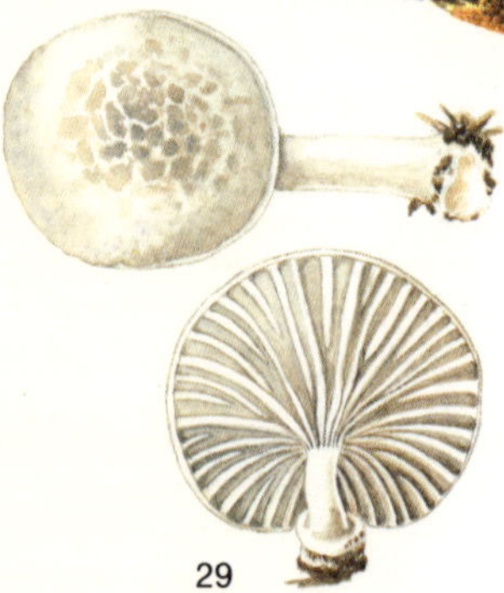

29

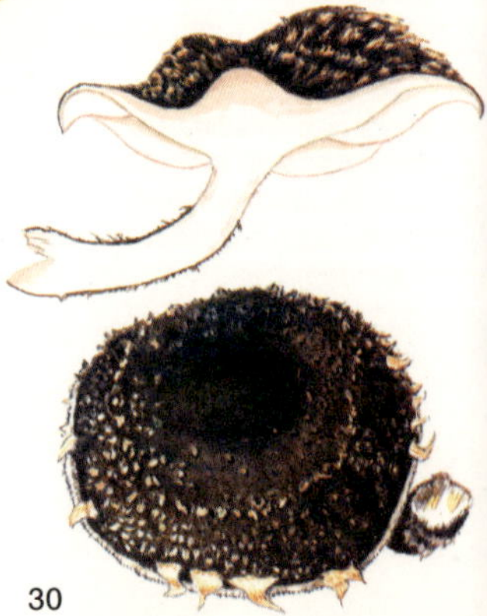

30

31

White-spored toadstools growing on wood

28. *Xeromphalina tenuipes* (Schw.) Smith (=*Crinipellis velutipes* Stevenson). A tough toadstool growing on dead trunks and fallen logs of hardwood trees. The cap is moist golden-yellow, the stem velvety ochre and the gills pale creamy or pink. It may be confused with *Flammulina velutipes*, which also has a velvety stem, but *Flammulina* has a sticky, gelatinous cap and its stem is brown grading to black at the base.

29. *Oudemansiella* sp. Has a translucent, whitish cap and an attachment pad at the foot of the stem. Also present in New Zealand is *Oudemansiella longipes* (St. Amans) Moser. It has a moist, greyish, wrinkled cap, thick white gills and a tall white stem with a scurfy surface. It also grows on or near wood.

30. "*Crinipellis novae-zelandiae*" Stevenson. The deep purple-brown cap is draped with rosy fawn fibrils from the veil. Parts of the same veil form shaggy scales at the cap edge, and cover the stem below the ring. The gills are tinted rose-lilac. This species has not yet been classified in its correct genus and it is a *Lentinula* related to the edible Japanese shiitake mushroom according to Professor K. Yokoyama.

31. *Tricholomopsis rutilans* (Fr.) Sing. *Edible*. An introduced species which grows on pine wood. The yellow cap is covered with a felt of maroon-purple fibrils, thicker and more dense in young specimens, more sparse and brown in age. The gills are bright yellow.

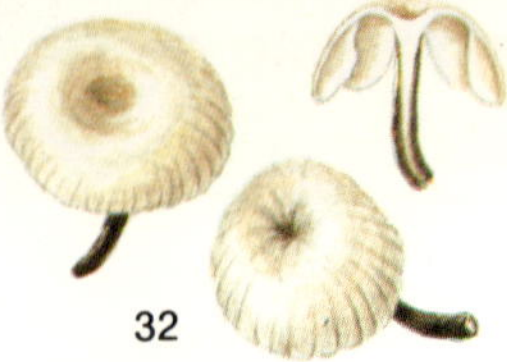
32

33

34

35

36

37

38

39

40

41

Small white-spored agarics

Marasmius is a genus of tough toadstools which shrivel under dry conditions and revive again when it rains. A common introduced species in lawns and fields is *M. oreades*, the fairy ring mushroom. Its cap, gills and tough stem are all creamy white, and it appears in autumn growing in the taller, greener grass just outside the bare zone where its mycelium has killed the grass.

32. *Marasmius* sp. Grows on wood.

33. *Marasmius atrocastaneus* Stevenson. Grows on soil or litter in mixed forest and beech forest.

34. *Crinipellis procera* Stevenson. Grows from sticks on the forest floor. It has twisted, horsehair stems covered with fine velvet, growing up to 12 cm high. An enlarged drawing of the cap is shown.

35. *Leucocoprinus fragilissimus* (Rav.) Pat. A delicate species growing on the ground in mixed forest.

36. *Mycena leaiana* (Berk.) Saccardo. Grows in clusters on wood. Its cap and stem are exceedingly sticky, and its colour changes from rich brown to pale grey as it ages. The gills are bright orange.

37. *Delicatula* sp. Parachute Mushroom. Some kinds have a stem, others are without one. The gills are widely spaced and may consist of shallow ridges.

38. *Mycena* sp. aff. *pura* (Pers. ex Fr.) Kummer.

39. *Mycena subviscosa* Stevenson. Belongs to subgenus Collopus, which has sticky stems.

40. *Mycena miniata* Stevenson. This mushroom is very conspicuous by its colour and gills with crimson margin. It belongs to subgenus Galactopus because the stem yields juice when broken. Found in mixed forest and beech forest.

41. *Mycena* sp.

There are many other different species of *Mycena* which look like fluted lampshades because of the radial lines or grooves near the cap margin.

Also common are: *Mycena veronicae* Stevenson in clusters on wood. It has a whitish cap, gills running down the stem, and the stem is covered with clear jelly. *Mycena veneta* Stevenson grows singly but gregariously on wood, has a blue-green cap fading to yellow-ochre and a white stem with a blue-green basal pad.

42

43

44

45

46

47

48

49

Wax-gills — Family Hygrophoraceae

The members of this family have a translucent, waxy texture particularly evident in the gills. Some are drab in colour, but many have bright and striking colours, and the colourful ones often come to the attention of naturalists. The introduced *Hygrocybe conica* Fr., with its pointed, conical, red cap and yellow stem, is seen on lawns and in gardens. It excites interest as it blackens in age or on handling. *Bertrandia astatogala* Heim is found in native vegetation, and resembles a small *Hygrocybe conica*, which blackens exceedingly quickly and has a network of black, raised fibrils on the cap and stem.

42. *Hygrophorus viridis* Stevenson. Belongs to subgenus *Gliophorus* with its cap and stem covered by clear jelly.

43. *Hygrophorus* sp. One of a number of bright tomato-red or crimson species found in mixed forest and *Leptospermum* scrubland.

44. *Gliophorus versicolor* Horak. This mushroom is slippery with an outer layer of jelly. The colouring changes from lilac-blue when young to faded pink. The stem is variously blue or pink with a yellow foot.

45. *Hygrophorus salmonipes* Stevenson. Grows in mixed forest and under *Leptospermum*.

46. *Hygrocybe procera* (Stev.) Hk.

47. *Hygrocybe striatolutea* Horak

48. *Hygrocybe luteovirens* Horak

49. *Hygrocybe rubro-carnosa* (Stev.) Hk. A handsome, red, velvety species in which the surface fades to yellowish with age but the brilliant crimson flesh persists throughout its life.

50

51

52

53

54

55

Russula — Family Russulaceae

Russula and *Lactarius* are set apart from all the other agarics by the brittle texture of their flesh. This can be tested by bending the stem, which will either snap or crumble like a piece of chalk. Under a microscope the reason for this brittleness can be seen, as the flesh contains clusters of spherical cells mixed among the more usual weft of fungal hyphae.

Russulas are conspicuous, as mycorrhizal associates with *Leptospermum* and *Nothofagus* as well as with some introduced trees. They are readily recognised by their thick, whitish gills set widely apart in a single series or sometimes forking. The spore print is white to pale beige (see p.27). Edibility of the New Zealand species is not known, but a number of northern species are used for food. Taste is used to classify the group as some have a hot, peppery flavour. *Caution*: Test by nibbling and do not swallow any.

50. *Russula* sp. Purple Russula. There are a number of similar looking species associated with *Leptospermum* and *Nothofagus*.

51. *Russula albolutescens* McNabb. White with ochre-yellow stains in age. Near the cap margin are radiating, bumpy ridges. Tastes mild and associates with *Leptospermum*.

52. *Russula* sp. Green Russula. Associates with *Nothofagus*.

53. *Russula miniata* McNabb. Associated with *Nothofagus* there are three rosy pink Russulas recognised by McNabb:
R. roseopileata McNabb. Tastes peppery, stem white.
R. vivida McNabb. Tastes mild, stem with pink tints, cap up to 4.5 cm diameter.
R. miniata McNabb. Tastes mild, stem white, cap up to 2.5 cm diameter.
Associated with *Leptospermum* is *R. pudorina* McNabb. Tastes horribly bitter, cap and stem both pink.

54. *Russula acrolamellata* McNabb. A yellow-brown species which stains darker brown on the gills and stem in age or when damaged. It has radiating bumpy ridges near the margin. Tastes peppery and associates with *Leptospermum*.

55. *Russula griseoviridis* McNabb. Has a pale grey-green outer membrane in the button stage. This cracks into persistent filmy patches as the cap extends. Tastes mild and associates with *Leptospermum*.

56

57

58

59

Lactarius — the milk caps — Family Russulaceae

Lactarius is very closely related to *Russula* in general appearance, in structure of the flesh and in mycorrhizal associations. It is distinguished from *Russula* by the oozing of latex or "milk" from any part which is cut or damaged. From this feature the common name of Milk Cap is derived. A slight bending of the cap or a light cut across the gill edges will produce drops of exudate in all but very old specimens.

When young, the cap of *Lactarius* or *Russula* is rounded, and during development it flattens out or becomes slightly depressed in the centre while the margin is either turned down or curled under. There is no trace of a veil in either genus.

56. *Lactarius sepiaceus* McNabb. A dark velvety milk cap associated with *Nothofagus*. It has creamy white gills and white milk, unchanging on exposure to air.

57. *Lactarius umerensis* McNabb. A small rubbery species up to 3.5 cm diameter, growing in scattered groups under *Nothofagus* or *Leptospermum*. Its milk is white and unchanging.

58. *Lactarius clarkei* Cleland. Named after Miss Clarke who painted watercolours of Australian toadstools, it is massive, up to 8 cm diameter, the cap velvety or suede-like. Its milk is white when fresh but dries leaving brown spots on the gills. The variety from under *Nothofagus* is bright orange with orange tinted flesh while the *Leptospermum* variety has more subdued colour and white flesh.

59. *Lactarius tawai* McNabb. Zoned Milk Cap. A large species which grows under *Nothofagus* and has a rather sticky, unchanging, creamy white milk.

60
61
62
63
64
65
66

Entoloma — toadstools with pink, angular spores.

Entoloma is a large genus containing toadstools of all sorts of shapes and colours. They agree in having a salmon-pink to brick-pink spore print (see p.27), and their spores are angular shaped when seen under a microscope. There is no veil present and the species are probably decomposers of humus and plant debris.

60. *Entoloma perzonatum* Horak. Zebra Toadstool. An easily recognised species which appears to be an inhabitant of open grassland, especially at high altitudes.

61. *Entoloma* sp. A stout, fleshy, lavender-grey species found under *Leptospermum* and in kauri forest.

62. *Entoloma peralbidum* Hk. White cap with a central umbilicus and gills decurrent (running down the stem). In mixed forest and under *Leptospermum*.

63. *Entoloma hochstetteri* (Reich.) Stevenson (=*Entoloma virescens* (Berk. & Curt.) Hk. if it is the same as the Japanese species). This is the incomparable sky-blue toadstool, always an exciting find whether or not you have seen it before. The colour can be paler or darker blue and fades with age. The cap too can be very pointed in shape or more rounded at the apex. There are several conical species with different colouring:

E. procerum Stev. Brown in colour.
E. psittacinum (Romagn.) Hk. Greenish-grey in colour.
E. canoconicum Hk. Grey with pale yellowish peak in colour.
Entoloma sp. Brick red in colour.

64. *Entoloma sulphureum* Hk. Yellow, with a faintly zoned cap, from mixed forest.

65. *Entoloma decolorans* Hk. From mixed forests.

66. *Entoloma haastii* Stev. A large sooty-blue species which appears to prefer *Nothofagus* habitats but is also found in other kinds of forest.

67
68
69
70

Other pink-spored toadstools

These species have in common a pale to dark salmon-pink spore print (see p. 27), but they are not closely related to one another, or to *Entoloma* on the previous page.

67. *Pluteus* sp. A large toadstool of grassy places or near sawdust. The gills are paper thin and very crowded, with a fine black line of pigmented cells along the margins.

68. *Rhodocybe* sp. This species has tough flesh and thick, shallow gills with wavy margins. The spore print is only faintly pink, and the spores have a finely irregular wall, detectable only under high magnification.

69. *Pluteus velutinornatus* Stevenson. Usually found as a solitary fruitbody on damp, spongy wood. It has a beautiful, dark, velvety texture to the cap and a bumpy surface. As it weathers, the velvet cells wear away, leaving a netted pattern on the cap. (See upper illustration showing an elderly specimen.)

70. *Volvariella speciosa* (Fr.) Singer. Grows in richly manured ground in gardens and waste places, and has probably been introduced into New Zealand. It has a membranous, cup-like volva, and this, together with other microscopic characters, suggests it is related to *Amanita*. The gills are white at first, then become pink with the spores. There are two forms, one with a snow-white, sticky cap, the other with a grey, fibrillar centre to the sticky cap. Be careful not to confuse this with the field mushroom.

71
72
73
74
75
76

Cortinarius — brown-spored curtain fungi

Cortinarius must be one of the largest genera of toadstools in New Zealand. It is readily recognised by the delicate cobwebby veil or cortina (Latin "little curtain") joining the cap margin to the stem in young fruitbodies. This veil stretches and tears as the cap expands, but a trace often persists at maturity as a bracelet of fine strands powdered with rusty or yellow-brown spore dust. The spores have rough walls when seen under the microscope.

Names of subgenera into which many agaric specialists would divide *Cortinarius* are mentioned in brackets.

71. *Cortinarius* sp. (=*Phlegmacium*). Recognised and never forgotten because of its smell of stale garlic.

72. *Cortinarius* sp. (=*Phlegmacium*). Blue and Green Curtain Fungus. The colours are washed out by rain, and in age the cap is patched with ochre-yellow, with the blue colour remaining only in the upper stem flesh.

73. *Cortinarius* sp. (=*Dermocybe*). Crimson and Gold Curtain Fungus. Very handsome in its regal colouring.

74. *Cortinarius* sp. (=*Dermocybe*). Canary yellow Curtain Fungus.

75. *Cortinarius* sp. (=*Cortinarius* in a restricted sense).

76. *Cortinarius* sp. (=*Myxacium*). Blue and Orange Slippery Curtain Fungus. Both the cap and the stem below the ring are covered in slimy, orange-tinted jelly. Underneath the jelly the toadstool has violet-blue colouring. The gills are lavender-blue when young, then become rusty brown as the spores mature.

77

78

79

80

81

82

83

More *Cortinarius* and other brown-spored agarics

Cortinarius is found as a mycorrhizal partner of both *Leptospermum* and *Nothofagus* but is not confined to growing with these species. It grows also in the beech forests of South America, Tasmania and New Guinea; and in other kinds of forests all over the world. Only a small selection of the numerous species is able to be pictured here and on Plate 15.

77. *Cuphocybe olivacea* Heim. A very scaly toadstool related to *Cortinarius*. Found only in beech forest.

78. *Cortinarius* sp. A white species with dry cap and stem and an unusual, pale olive-greenish spore print.

79. *Cortinarius* sp. (=*Hydrocybe*).

80. *Descolea gunnii* (Berk.) Horak. Has a membranous ring with radial stripes on the upper surface. Found with *Nothofagus* or *Leptospermum*.

81. *Inocybe* sp. This genus has radially fibrillose to scaly caps, a dull greyish-brown spore print, and is found growing in soil.

82. *Inocybe* sp. (=*Astrosporina*).

83. *Inocybe strobilomyces* Horak. A large species with colouring that makes it difficult to see against the background of litter on the floor of the beech forest. The gills are pale lemon darkening to greyish. Overseas workers have reported several poisonous species of *Cortinarius*, which makes it unwise to try any of them for food. *Inocybe* is also a genus well known for very poisonous species, some fatal.

84

85

86

87

88

Brown-spored agarics growing on wood

84. *Pluteolus muscicola* (Stev.) Hk. A delicate species, at first glance like a *Pluteus*, but differing in that the gills and spores are pale orange not pink. The leopardskin pattern on the cap is made by a number of dark coloured depressions underneath the clear surface jelly. There are several plain yellow or whitish cousins of this bush species which come up on lawns and grass verges on dewy autumn mornings only to shrivel and disappear by midday. They are species of *Bolbitius*.

85. *Gymnopilus junonius* (Fr.) Orton. Forms huge clusters of vivid yellow toadstools on wood. *Gymnopilus* has a bitter taste and is not edible.

86. *Pholiota* sp. aff. *adiposa* (Fr.) Kummer. Has lemon coloured, very glutinous caps covered with brown scales floating in the gluten and scaly patches also covering the stem below the ring. The species is found often on lacebark (*Hoheria*) wood.

87. *Galerina* sp. This *Galerina* has a pointed cap, but there are others with a smoothly convex shape.

88. *Agrocybe* sp. aff.* *cylindracea* (DC ex Fr.) Maire. This can be of massive size (the cap to 20 cm diameter) in gigantic clusters on wood, or arising from a wound cleft in a living tree trunk. It may produce fruitbodies regularly each season from the same site on an infected tree. The dark chocolate spore print often falls on the cap below. The ring is full and shows longitudinal stripes from contact with the gills, but it soon collapses into a brown smudge. In dry weather the cap surface may break into flat areas separated by cracks. If this is the same species as *A. cylindracea* from Europe then it should be edible.

89
90
91
92
93

Coprinus — the ink caps

These agarics have black spores, and in many the cap is tall and narrow with crowded gills. The spores near the cap margin are the first to ripen and these are shed first. The aging process is accompanied by a darkening of the cap, and soon after the spores are dispersed, that part of the cap dissolves into an inky fluid. The process of self-destruction gradually moves upwards from the margin towards the apex until at last only the stem remains, stained with black.

89. *Coprinus* sp. aff. *disseminatus* (Pers. ex Fr.) S. F. Gray. Sociable *Coprinus* is not as melting as the other ink caps. It is found in droves, huge numbers covering a dead log. Young fruitbodies are pale creamy or yellowish with radial stripes and they later darken to grey with a yellow central spot.

90. *Coprinus* sp. A solitary *Coprinus* in which the cap curls upwards and becomes almost transparent as the spores are shed.

91. *Coprinus* sp.

92. *Coprinus comatus* (Mull ex. Fr.) S. F. Gray. Shaggy Ink Cap, Lawyer's Wig. *Edible*. The veil is sometimes visible as a movable ring on the stem. Its shaggy scales, white colour with a fawn apex, and cap which darkens from grey to black before liquefying are unmistakable. Remove stem and use cap at the young stage, before blackening occurs. Must be collected and eaten the same day.

93. *Coprinus atramentarius* (Bull. ex Fr.) Fr. Ink Cap. *Edible. Caution*: do not consume alcohol at the same meal as this ink cap, because the combination causes illness. Found in clusters growing from wood, it sometimes appears from soil but is then associated with dead roots or buried wood. The mark of the veil, placed low on the stem, identifies this species.

Also associated with wood and growing in clusters is *Coprinus micaceus* (Bull. ex. Fr.) Fr., which has a cream or fawn cap with glistening particles over the surface. It becomes a dark brown colour before dissolution.

Plate 18

94
95
96
97
98
99

More toadstools with black spores

94. *Stropharia aurantiaca* (Cooke) Orton. This is a very conspicuous tomato-red toadstool of rich soil in gardens, favouring places with sawdust or rotting wood. *Stropharia* differs from *Agaricus* in having the gills joined to the stem.

95. *Stropharia aeruginosa* (Curt. ex Fr.) Quel. Verdigris Agaric. Another strikingly coloured *Stropharia* of waste places. Its brilliant blue-green pigment washes out in rain leaving the cap partly or wholly ochre-yellow. It has a strange smell and is reputed to be poisonous.

96. *Psathyrella* sp. Bears some resemblance to an ink cap, with its thin cap and fragile white stem; but *Psathyrella* is not self-digesting. This species grows in mixed forest. Several other species with pale fawn or brown caps commonly appear in gardens, especially from decaying roots where a tree has been cut down.

97. *Panaeolus* sp. On rich pasture or waste ground. Some species of *Panaeolus* grow on dung pats. They have black mottled gills and tall slim stems. The cap is usually not striate.

98. *Lacrymaria velutina* (Pers. ex Fr.) Konrad & Maubl. Weeping Widow. Cap fawn with silky radial hairs. Gills mottled black from the spores and oozing watery drops in wet weather. Grows on rich pasture or waste ground.

99. *Naematoloma* sp. New Zealand Sulphur Tuft Agaric. This species is common on fallen wood and old stumps both in gardens and in undisturbed forests. It occurs in drifts and clusters of many individuals at a time. Sometimes it is yellow, or it may have red-brown colouring in the centre of the cap. The gills are dull yellow-green, becoming finally black with the maturing spores.

100
101
102
103
104
105
106

Secotiums or pouch fungi

These fungi are named after the genus *Secotium* into which they were classified before modern study enabled us to understand their diverse relationships. In structure they are like agarics which develop as far as the button stage, then the cap grows into a closed pouch which does not open to release spores. Often the gills become a labyrinth of chambers surrounding the stem, and if the stem is not well developed we have a tuberous fungus that is essentially a gasteromycete.

It is interesting to speculate why members of different agaric families have the secotioid form in New Zealand. *Weraroa* appears to be related to the black-spored *Stropharia; Thaxterogaster* to *Cortinarius*, and *Notholepiota* to *Lepiota*. Overseas, secotioid fungi are eaten and dispersed by small mammals, and some live very successfully in deserts. These explanations do not hold here, and it may be that the closed cap is to protect the spores from our wet climate, holding them until they are dispersed by fungus eaters like slugs or insects, or birds (like kiwis) which fossick for insects.

100. *Weraroa erythrocephala* (Tul.) Sing. & Smith. Red Pouch Fungus. Grows in mixed forest.

101. *Weraroa virescens* (Massee) Sing. & Smith. Pale Blue Pouch Fungus.

102. *Tympanella galanthina* (Cke. & Mass.) Horak. Has a white woolly cap and brown gills, but the cap rarely opens and spores are not released into the air.

103. *Thaxterogaster aurantiacum* Hk. Has a layer of bumpy jelly on the outside and a very reduced stem. It associates with *Leptospermum* and *Nothofagus* and grows in the soil, sometimes buried well below the surface.

104. *Thaxterogaster epiphaeum* Hk. Found under *Nothofagus*.

105. *Thaxterogaster porphyreum* (Cunn.) Hk. Violet Pouch Fungus. A well known fungus of *Nothofagus* forests.

106. *Notholepiota areolata* (Cunn.) Hk. Associated with *Leptospermum*.

Plate 20

107
108
109
110

Boletes and their gilled relatives — Family Boletaceae and Paxillaceae

Boletes have fruitbodies like agarics in shape and texture, but their spores are produced inside pores or tubes, not on gills. The *Paxillus* fungi have spores like the boletes and fruitbodies which are depressed in the centre to almost funnel-shaped, with the gills often forking and running down into the stem. In both, the pores or gills are easily peeled away from the cap.

107. *Boletus* (=*Suillus*) *granulatus* L. ex Fr. Slippery Jack, Sticky Bun Bolete. *Edible*. This common introduced fungus forms mycorrhizas with pine. It is variable in colour, with a pale brown to red-brown cap, sometimes mottled with deep brown or olive green. The sticky skin peels easily to reveal pale yellow flesh. The pores are lemon to deep brassy yellow, and the upper stem bears glandular dots which ooze latex. Prepare for cooking by peeling off the skin and tube layers. Taste each cap in case you have collected the sharp flavoured *B. acerbus*.

108. *Paxillus* sp. A native species found on lakeside flats close to *Nothofagus* forest.

109. *Boletus piperatus* Bull. ex Fr. Peppery Bolete. *Not edible* because of its strong taste.

110. *Paxillus nothofagi* McNabb. Forms huge fruitbodies up to 16 cm across the cap, growing in clusters in the soil. *Edibility not known*.

Two other common introduced boletes are edible. *Boletus* (=*Suillus*) *luteus* is like *B. granulatus* but has a membranous veil leaving a dark ring on the stem. It is an associate of pine. *Boletus* (=*Leccinum*) *scaber*, which grows under silver birch, has a pale fawn cap, whitish pores and a pale stem with patterns of black, scurfy granules arranged in more or less vertical rows. This birch bolete is eaten in Italy where it is known as "porcinello".

Another mushroom associated with silver birch is the introduced *Paxillus involutus*. In shape it is like *P. nothofagi* and when young it has a pale fawn or pink-fawn velvety cap, with the margin tightly rolled under and the gills forming a network where they run into the stem. As it matures the cap becomes brown, and all parts stain brown where they are touched.

Plate 21

111
112
113
114
115

Boletes native to New Zealand — Family Boletaceae and Strobilomycetaceae

111. *Porphyrellus novae-zelandiae*. McNabb. A very attractive looking species with its pepper-patterned cap and lilac-tinted pore surface. Its stem is veined with raised ridges. It grows with *Nothofagus* and in mixed forest containing *Leptospermum*, where it often appears on the basal root buttresses of tree ferns.

112. *Porphyrellus niveus* (Stevenson) McNabb. Has raised ridges on the stem like those of *P. novae-zelandiae*, but the whole plant is white with a sticky surface to the cap. Associates with *Leptospermum, Nothofagus* or mixed forest.

113. *Porphyrellus brunneus* McNabb. This species has a dry, cocoa coloured cap and a pale creamy pore surface, bruising deep blue and finally brown. It associates with *Leptospermum*.

114. *Porphyrellus viscidus* McNabb. Has a sticky, dark brown to purple-brown cap and unchanging lemon-yellow pore surface. Associates with *Leptospermum*.

115. *Tylopilus formosus* Stevenson. Dark Velvet Bolete. This is a large species with a velvety cap and stem, the pore surface being dingy lilac-pink and the flesh white. It associates with either *Leptospermum* or *Nothofagus*.

Porphyrellus belongs to the family Strobilomycetaceae and *Tylopilus* to the family Boletaceae but in fact the two genera are very closely related. Their edibility is not known but *P. novae-zelandiae* and *T. formosus* both have an exceedingly bitter taste.

116

117

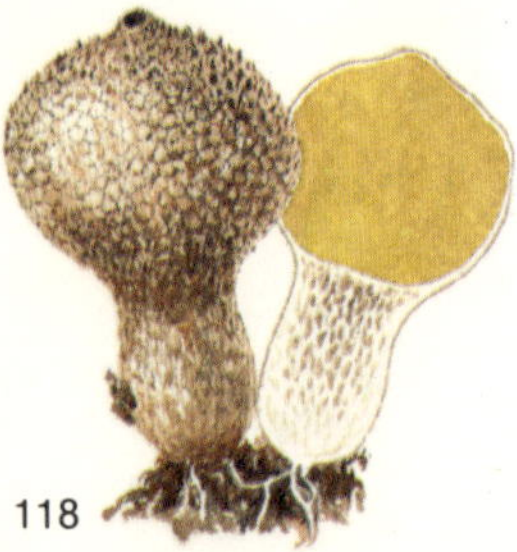

118

119

Earthballs and puffballs — Family Sclerodermataceae and Lycoperdaceae

116. *Scleroderma* sp. Earthball, Stone Fungus. *Not edible*, bitter tasting. These are hard lumps containing a solid mass of purplish spores. At maturity the purple spore powder is released by the skin flaking from the top to form a ragged crater with lobes that curl outwards. After the spores are gone the empty skin may be mistaken for an earthstar fungus. Another member of this family, *Pisolithus tinctorius* (Pers.) Coker & Couch, is found with pines or manuka on the volcanic plateau. It is a yellow earthball which bruises black and stains the fingers yellow. Inside, the spore tissue is divided up into compartments which turn black as they ripen, and the spores are released by the gradual downwards weathering of the fruitbody.

117. *Calvatia utriformis* (Bull. per Pers.) Jaap. Mosaic Puffball. Grows in lawn or pasture.

118. *Lycoperdon* sp. Common Puffball. Grows in soil on the forest floor or in pasture. Entirely white when young, it darkens towards maturity.

119. *Calvatia cyathiformis* (Bosc) Morgan. Lilac Puffball. Grows in lawns and pastures.

Puffballs differ from earthballs in that they have a soft, fleshy texture and remain white inside until they reach nearly full size. The spore compartment may take up all of the interior or there may be a sterile base. The spores themselves are entangled amongst brown "cotton wool" which ensures that they sift out sparingly over a period of time. In *Calvatia* spores are released by irregular cracking of the wall, but in *Lycoperdon* there is an apical pore, and the impact of raindrops or the tap of a finger causes the spores to puff out.

Puffballs are edible, but they must be fresh and white right through. Cut them in half and look for traces of colour which warn you they are too old for eating. This test will also sort out the earthballs, which have purple colouring very early, and any button toadstools will be recognised by their developing gills. *Calvatia gigantea* (Batsch ex Pers.) Lloyd is the smooth Giant White Puffball, the size of a football or larger. It is probably the best species for eating as it can be recognised easily, and one fruitbody is enough for a meal.

Plate 23

120

121

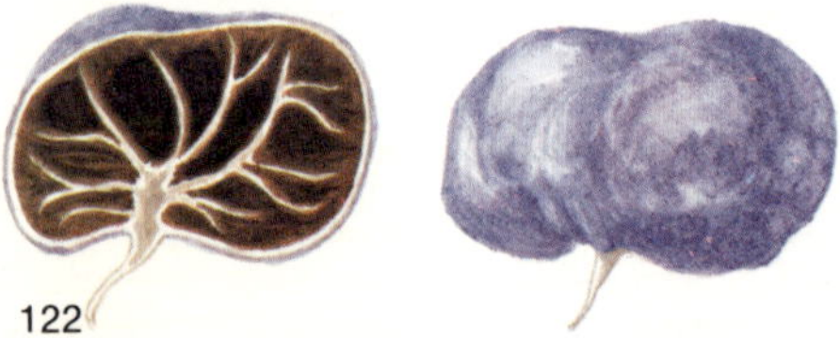
122

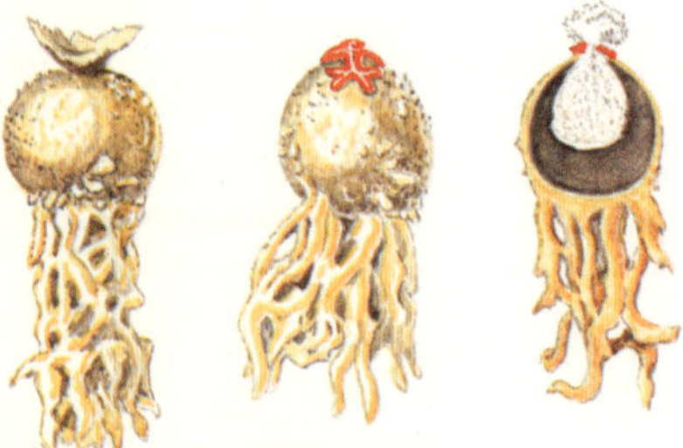
123

124

125

126

Tuberous fungi, puffballs and earthstars — Gasteromycetes

120. *Gautieria novae-zelandiae* Cunningham. A white potato fungus half buried in soil under *Nothofagus*. In age it stains dark metallic green and it has an unpleasant smell.

121. *Melanogaster* sp. Grows under introduced trees (oak). It is red-brown with surface cords (rhizomorphs) and a black and white interior.

122. *Hysterangium sclerodermum* (Cooke) Cunn. Violet Potato Fungus is very like the violet *Thaxterogaster* but is not stalked. The white veins running through the spore tissue are gathered at the base into a root-like cord. Associates with *Nothofagus*.

Another tuberous fungus seen under introduced pines is *Rhizopogon*. It is pale at first then fawn, with surface cords like those of *Melanogaster*. The interior of *Rhizopogon* is brownish and spongy. It is an important mycorrhizal partner of *Pinus*.

123. *Calostoma rodwayi* Lloyd. A small puffball raised up on a pseudostem of anastomosing gelatinous strands. A piece of the outer scurfy layer peels off from the apex and exposes the red, lobed mouth. Then the spore sac inside gradually shrinks and pushes out the white spores. *Calostoma* associates with *Nothofagus*.

124. *Bovista plumbea* Pers. Leaden Puffball. Grows in open pasture. The outermost coat is papery and peels off in irregular patches to expose the inner leaden-grey wall and apical pore. The puffball often breaks away from its attachment and rolls along in the wind, scattering spores as it goes.

125. *Geastrum pectinatum* Pers. Beaked Earthstar. The central spore case is stalked and has a tall, beaked mouth. The outside of the rays is always entangled in soil and leaf litter.

126. *Geastrum velutinum* Morgan. Velvet Earthstar. In this species the central puffball lacks a stalk and remains in a bowl formed by the outer rays. The external surface of the rays is velvety and free from soil and debris.

Earthstars are a kind of puffball with a layered coat. The outermost layer splits and peels back into petal-like rays, thus clearing a space around the mouth. In some species the rays curl right back and lift the central puffball high into the air for more efficient spore dispersal.

Plate 24

127

128

129

130

131

132

Stinkhorns — Family Phallaceae

This family is one of the most strange in the fungus kingdom. Its members attract flies for spore dispersal, and to this end they mimic the appearance and smell of dung or carrion.

The young fruitbody is an "egg" which looks like a small puffball. It has a two-layered wall, the outer skin being lined with a thick layer of jelly. Inside the wall are enclosed two further tissues: a mass of olive-brown spores or gleba, and a receptacle folded up and compressed into a small space. Under moist conditions the receptacle absorbs water and swells, first bursting out of the egg wall and then unfolding to carry up with it the now liquefied spore mucus. Soon after hatching the stench from the gleba is detected by flies which come to feed on it, and when they leave they carry away spores as both inside and outside passengers, to be deposited later far away from the parent fungus.

127. *Aseroe rubra* La Bill. ex Fr. Flower Fungus. Receptacle like a flower or sea anenome, with a white stem crowned by a circle of horizontal, spreading arms forked towards the tip. The gleba is borne on a ring formed by the bases of the arms.

128. *Clathrus chrysomycelinus* Moeller. Golden Basket Fungus. Receptacle with a basal white stem (hidden inside the egg) which divides above into a golden-orange network. The gleba is collected into droplets borne on the inside of the lattice at the points where its branches intersect.

129. *Mutinus bambusinus* (Zoll.) Fischer. Bamboo Stinkhorn. Receptacle a narrow horn tapering at both base and apex. Gleba covers a red area towards the apex, which is perforated.

130. *Mutinus* sp. Yellow stinkhorn. This may be a colour variant of 119. It is found in mixed forest, whereas *M. bambusinus* is often associated with bamboos.

131. *Ileodictyon cibarius* Tul. Basket Fungus, Lattice Fungus. The receptacle can be very large, up to 16 cm across. It is a spherical white net with the gleba smeared all over the inside of the meshes. The basket can break free and be rolled away some distance from where it hatched. The Maori explained the miraculously sudden appearance of these fungi by saying they were the droppings of ghosts.

132. *Pseudocolus javanicus* (Penzig) Lloyd. Receptacle of three or four slim arms on a small stem, the arms being lightly joined at their tips on hatching but later becoming free. Pink and red arms are also found.

133. *Anthurus archeri* (Berk.) Fischer, Giant Stinkhorn. (See frontispiece.) The giant stinkhorn has a pink coloured "egg" and may be up to 20 cm tall. Receptacle a circle of four to ten salmon to scarlet arms borne on a hollow stem. The arms are single and taper down gradually into the stem without an abrupt distinction between the two regions. The gleba is spread thinly over the inside of the arms, which are joined at their tips when they first emerge.

134

135

136

137

138

139

140

141

A. Birdsnest fungi — Family Nidulariaceae

These are a very specialised family of gasteromycetes. They have small fruitbodies which are completely closed at first, but then the top crumbles away to leave a cup containing peridioles. The peridioles are pellets of compressed spore-bearing tissue and the whole arrangement looks like a tiny nest containing eggs. It is a splash cup, and when a raindrop scores a hit, the peridioles (eggs) are propelled into the air.

134. *Crucibulum laeve* (DC.) Kambly. Common Birdsnest Fungus. The cup of this species is yellow-brown and felted on the outside. The peridioles are flattened, whitish, flying saucers with an elastic cord of complex structure attaching them to the nest. When they are splashed out of the nest, the sticky knob at the free end of the cord catches on the surrounding vegetation, and the backlash from the unravelled cord is enough to coil it firmly around the nearest support.

135. *Nidula candida* (Peck) White (or *N. niveo-tomentosa* (Henn.) Lloyd). White Woolly Birdsnest. The cup is white and woolly outside, orange inside and the peridioles are like shiny, oval, red-brown seeds, without an attachment cord. The entire surface of this peridiole is sticky and it adheres to anything it lands on.

B. Coral fungi and fairy clubs — Family Clavariaceae in a wide sense

These fungi used to be classified together because of their similar shapes. We now know that they are an artificial grouping of species which have a very different internal structure; but it is convenient to deal with them together here.

136. *Clavariopsis* sp. aff. miniata (Berk.) Corner.

137. *Clavulina cavipes* Corner.

138. *Clavicorona colensoi* (Berk.) Corner. Has regular, candelabra-type branching and is always found growing on wood.

139. *Ramaria* sp. Has a large base and poorly developed branches.

140. *Aphelaria dendroides* (Jungh.) Corner. This coral fungus has a tough, rubbery texture.

141. *Ramaria* sp. aff. formosa (Fr.) Quel. *Poisonous*. Spore print yellow-brown and spores rough. Bruises brown.

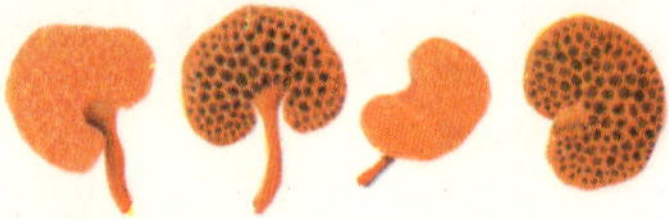

142

144

143

146

145

148

147

149

Woody pore fungi — Family Polyporaceae

Polypores have a more complex microstructure than agarics, their flesh being composed of several kinds of hyphae. The strengthening and binding hyphae account for the tough texture and long life of these fruitbodies. The family is one of convenience, members resembling one another only in their general appearance and in growing on wood.

142. *Favolaschia calocera* Heim. Orange Pore Fungus. An introduced tropical species, it is now spread abundantly throughout the North Island. The stem is short and attached laterally to one side of the rubbery cap. Handling this fungus stains the fingers orange.

143. *Ganoderma australis* (Fr.) Pat. Perennial Bracket, Artist's Conk. A large, tough bracket, almost impossible to remove from its host except with an axe. Its upper surface has a dense crust, the flesh is brown and the pore surface is pale or dingy at first, aging and bruising brown. Pictures scratched on the pore surface will remain visible on drying.

144. *Grifola berkeleyi* (Fr.) Murrill (=*Bondarzewia*). This grows in huge tufts or rosettes aggregated together on a common base at the foot of dead trees. The pore surface is whitish, and the fruitbody exudes pale latex when broken.

145. *Polyporus arcularius* (Batsch) Fr. A small toadstool-shaped polypore.

146. *Coltricia oblectans* (Berk.) Cunn. Silky brown zoned fans or umbrella-shaped fungi with pores underneath, growing in soil or moss.

147. *Pycnoporus cinnabarinus* (Jacq. ex Fr.) Karst. Orange Bracket. This fungus is the colour of cinnabar, an ore of mercury. Weathered specimens may be pale in colour, but retain bands of orange and white in the flesh.

148. *Coriolus zonatus* (Nees ex Fr.) Quel. Rainbow Bracket. Thin with white pores and flesh and bands of fawn and red-brown on the upper surface, some furry and some smooth. Because new growth takes place along the exposed outer edge, the character of the surface is altered by fluctuations in the weather, and this causes the zonation. Brackets often grow fast enough to surround and enclose pieces of vegetation near to them. *Coriolus versicolor* is similar but thicker, and includes bands of dark blue and grey as well as brown.

149. *Piptoporus portentosus* (Berk.) Cunn. A hoof- or bell-shaped bracket with very thick, slippery, white flesh which is tough to cut. The pores are shallow, with a white lower surface.

150

151

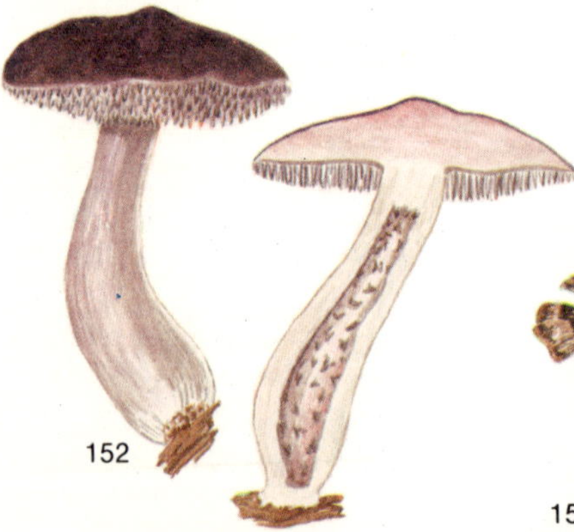
152

153

154

155

Spine or tooth fungi — Family Hydnaceae

This is another artificial grouping of fungi which superficially resemble one another; in this case they all have the spore-bearing layer on hanging teeth or spines. Like the corals in Plate 26, the polypores in Plate 27 and the crust fungi in Plate 29 they are being subjected to modern methods of study, and the older classification which is used here is certain to be discarded in favour of one based on new insights into the true relationships of the species.

150. *Hydnum crocidens* var. *crocidens* McNabb. This closely resembles the "wood hedgehog" of the north temperate regions. It tends to stain gold-orange on bruising or in age. It should be *edible* like the northern species. It associates with *Leptospermum* and *Nothofagus*.

151. *Hydnum crocidens* var. *badius* McNabb. A chestnut-brown variety of the above.

152. *Sarcodon carbonarius* Maas Geesteranus. Purple Hedgehog. The dull, dark purple-black cap is easy to overlook when collecting. The spines and flesh are drab lilac or pinkish when damaged or cut, and the stem hollow. The whole fungus dries black.

153. *Phellodon sinclairii* (Berk.) Cunn. A tough fan-, shelf- or rosette-shaped fungus, purple-black with shining zones above. Underneath, the fertile surface is white to pale, dull lavender with hanging teeth. As with 152, the whole fungus dries brittle and carbonaceous in age. It has a distinct smell of aniseed and associates with *Nothofagus*.

154. *Hericium coralloides* (Scop. ex Fr.) S. F. Gray. Fungus Icicles, Stalactite Fungus. *Edible*. A white, much-branched fungus growing from decaying wood. The tips of the branches hang downwards and the fungus may take on pink tints in age.

155. *Phellodon nothofagi* McNabb. Very similar to 153 except for its colour. It also associates with *Nothofagus*.

156

157

158

159

160

161

162

Thelephores, crust fungi and smooth, thin bracket fungi

156. *Thelephora terrestris* Fr. Thick shelves or brackets growing in soil, often encrusting sticks or roots, very variable in shape. Upper surface irregular and purple-brown, lower fertile surface irregular to smooth, pale purplish with a white margin. Spore print purple-brown like that of an *Agaricus*. Sometimes the margin is fringed and cut instead of smooth, as figured. Associates with pine and larch.

157. *Podoscypha petalodes* ssp. *floriformis* (Bres.) Reid. An eye-catching, ruffled, wineglass-shaped fungus, on close inspection seen to be formed from a circular fan with its margins overlapped. The inside (upper) surface is silky and zoned or banded in shades of fawn-pink and brown. The outer pale surface bears the spores. Sometimes this fungus forms a longitudinal ruffle along wood or buried wood, but it can be recognised by the short tough stems attaching it to the substratum.

158. *Stereum complicatum* Fr. Small rosettes on wood, attached at the centre, free and ruffled at the margins.

159. *Aleurodiscus ochraceo-flavus* Lloyd. Forms patches on *Leptospermum* wood. It was named from dried specimens, for when fresh the bright pink colour of the fertile surface is quite outstanding. It is attached over a broad base, with a free, white, woolly margin.

160. *Stereopsis hiscens* (Berk. & Rav.) Reid. Small, erect, branched fans growing from soil. They are white and become purple-brown on the spore-bearing surface.

161. *Stereum fasciatum* (Schw.) Fr. A thin bracket, found especially on tawa and taraire (*Beilschmiedia*) wood. The fans can be up to 7 cm wide, with a narrow supporting emplacement fixed to the wood on the underside. The upper illustration shows the upper side, the lower one the smooth, fertile underside.

162. *Stereum vellereum* Berk. Small, conical or saucer-shaped fungi attached to wood (*Nothofagus*) by a small point at the apex or centre.

There are a great many fungi which grow on dead wood and can be described as paint-splash fungi or crust fungi. They are difficult to distinguish from the encrusting lichens which live with them on the same branches.

163

164

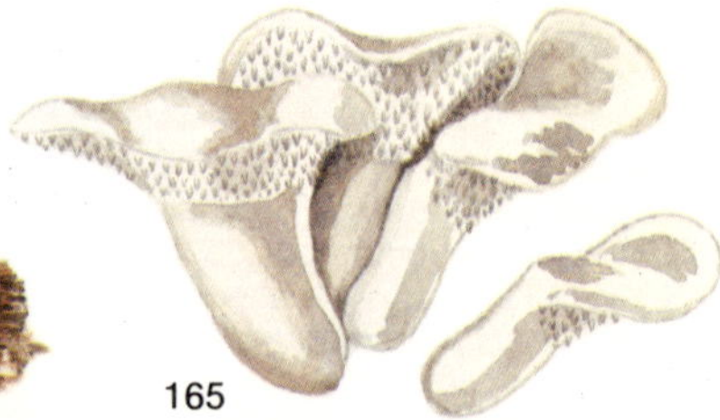

165

166

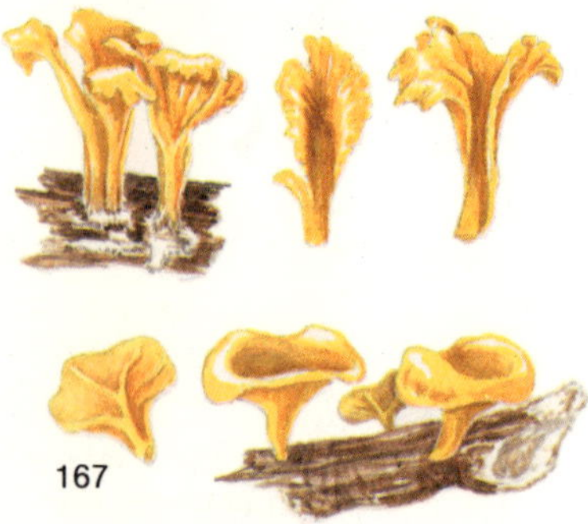

167

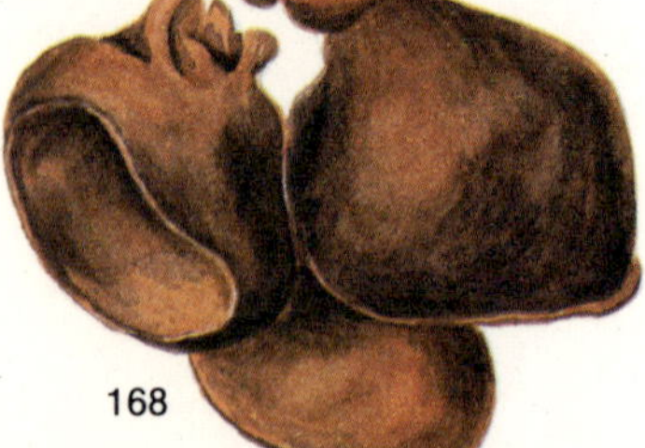

168

169

Jelly fungi — of three different families

These fungi all grow on wood and have a gelatinous texture, sometimes soft and jelly-like, sometimes tough and rubbery. They are all basidiomycetes, but they differ considerably from other basidiomycetes and from each other in the microscopic structure of their basidia.

163. *Tremella fuciformis* Berk. Tremellaceae. The soft white lobes of this fungus are gathered into a very small point of attachment to the wood. They collapse quickly in drought, but can be viewed in their proper shape by placing them in water. There is a yellow species, *Tremella lutescens*.

164. *Exidia glandulosa* Fr. Witches' Butter. Tremellaceae. This forms extended, irregularly lobed, pale grey or fawn translucent patches. The convoluted, fertile upper surface is beset with tiny, peg-like warts.

165. *Pseudohydnum gelatinosum* (Scop. ex Fr.) Karst. Tremellaceae. Jelly Hedgehog Fungus. A translucent fungus but firm in texture. The stalked fans may be solitary or clustered, and the fertile surface underneath consists of hanging spines.

166. *Calocera* sp. Dacrymycetaceae. Has small, creamy or pale flesh-coloured, pointed horns about 1 cm high growing together in a group. They may be curved when they have turned to grow vertically on a sloping log.

167. *Dacryopinax spathularia* (Schw.) Martin. Dacrymycetaceae. These are flattened spatula-shaped fungi with ruffled margins. They are attached very firmly to the host wood.

168. *Heterotextus miltinus* (Berk.) McNabb (= *Guepiniopsis*). Dacrymycetaceae. Jelly Buttons. These are stud-shaped, with a tendency to form ridges or angles on the outside.

169. *Auricularia polytricha* (Mont.) Saccardo. Auriculariaceae. Ear Fungus, Wood Ears. *Edible*. Ear fungus is abundant on rotting wood in lowland forests. It is brown to red-brown with a softly velvety exterior when young. Older fruitbodies often dry out and have a pale grey, furry exterior, while the smooth fertile surface within becomes varying shades of deep brown or purple and has a pale bloom from the spores.

The tough rubbery texture is not attractive to European tastes, but when prepared by soaking in water, and shred cut, ear fungus is a tasty addition to stir-fried Chinese dishes. New Zealand once exported large quantities of dried ear fungus to the East.

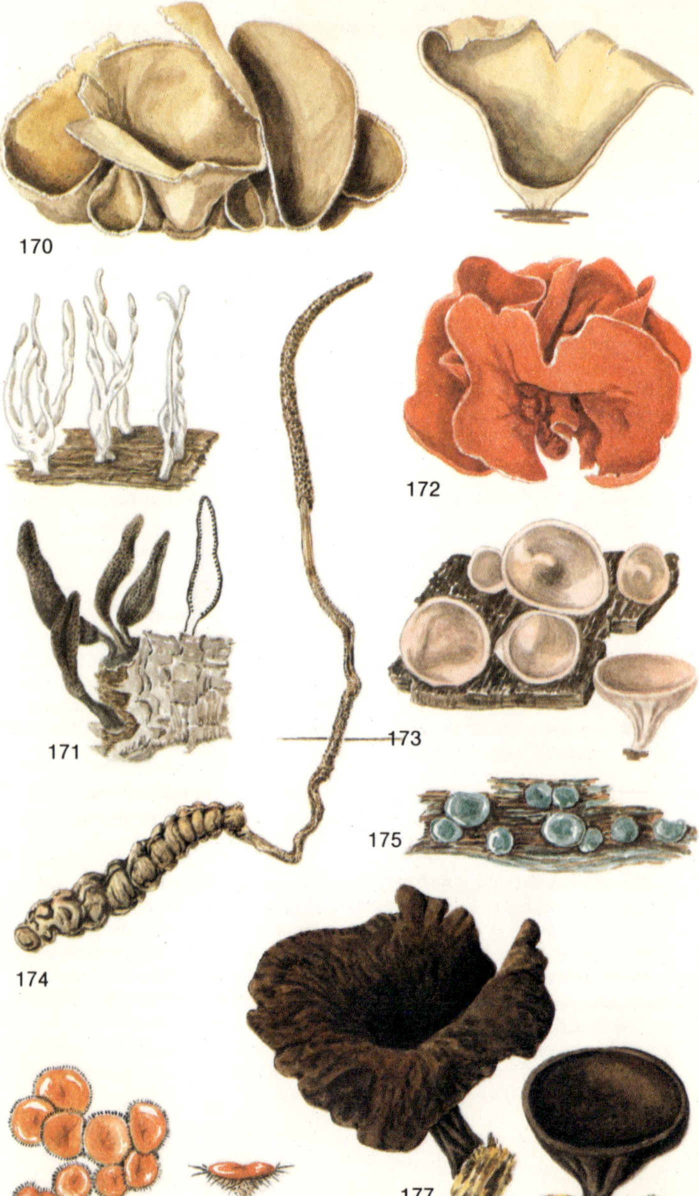
170
171
172
173
174
175
176
177

Cup fungi and other Ascomycetes

170. *Peziza vesiculosa* Bull. ex St Amans. Fawn Cup Fungus. Common around compost heaps and on cardboard and sacks left lying on the ground. The cups often grow packed together and then become compressed to fit into the space available to them.

171. *Xylaria polymorpha* (Pers. ex Mer.) Grev. Dead Man's Fingers. This species may be found producing white asexual spores (conidia) as in the upper illustration. Later it produces black fruitbodies with a bumpy surface as in the lower illustration. The bumps have a central pin hole which leads into a flask-shaped cavity (perithecium) containing the ascospores. If the fruitbody is cut you can see the row of black perithecia embedded in the white flesh.

172. *Aleuria aurantia* (Pers. ex Gray) Fuckel. Orange Peel Fungus. *Edible*. It usually grows on disturbed ground and the outer surface is velvety and paler than the orange, fertile surface inside.

173. *Cookeina colensoi* (Berk.) Le Gal. A pale pink cup fungus attached to wood by a small pleated stem. It is often found on fallen tawa, a hardwood.

174. *Cordyceps robertsii* Hooker. Vegetable Caterpillar. The illustration shows the mummified body of a soil-dwelling moth caterpillar killed by the fungus. When the fungus has used up all the nutriment provided by the insect tissues, it sends up a brown stalk which is the only part that appears above ground. The spore stalk is thickened at the apex like a tiny bulrush, being covered by globules which are the perithecia producing the ascospores.

175. *Chlorosplenium aeruginosum* (Leder ex Fr.) de Not. Verdigris Stud Fungus. The mycelium of the fungus stains the host wood green.

176. *Scutellinia colensoi* (Massee) Le Gal. Eyelash Elf Cup. These small, orange-red fungi have dark, bristly hairs on the underside which project as a black eyelash fringe around the margin.

177. *Plectania campylospora* (Berk.) Nannf. Black Cup Fungus. The dark cups of this species may be up to 8 cm across. Old specimens often expand into more irregular shapes. Black cup fungi are common on wood of *Leptospermum* and *Melicytus.*

178

179

180

181

182

183

184

More Ascomycetes

178. *Cyttaria gunnii* Berk. Beech Strawberry. *Edible*. The "strawberries" appear in spring, growing from the irregular woody galls their parasitic hyphae have induced on the branches of silver beech trees. After a few years the gall may girdle the branch and kill it. The fruitbodies are misnamed, for they have neither the rosy colour nor the succulent taste of strawberries.

179. *Trichoglossum hirsutum* (Pers. ex Fr.) Boudier. Earth Tongue. Grows in soil or amongst moss. This species has a dull suede-like texture while another earth tongue, *Geoglossum*, is shining black.

180. *Leotia lubrica* Pers. Slippery Lizard Tuft. This has such a slimy stem that it is almost impossible to pull it out of the ground. The irregular, dark green or olive-brown head produces the ascospores.

181. *Daldinia concentrica* (Bolton ex Fr.) de Not. Charcoal Fungus or King Alfred's Cakes. Brownish to black lumps growing on wood. The ascospores are produced in small cavities (perithecia) in the outermost layer of the fruitbody. A fruitbody left overnight in a closed container will shed masses of sooty black ascospores. This fungus has black flesh marked with concentric zones.

182. *Paurocotylis pila* Berk. From the outside this looks like the red pouch fungus, but it is without a stem and the tissue inside is whitish.

183. *Morchella esculenta* Pers. ex St Amans. Morel. *Edible*. Morels have fawn, pitted heads, and are hollow throughout both the cap and the stem. They must be carefully compared with the brain mushroom (no. 184) before they are eaten. Morels appear on the pumice soils of the volcanic plateau.

184. *Gyromitra esculenta* (Pers.) Fr. Brain Mushroom, Lorchel. *Poisonous*. The wavy, convoluted head with some irregular, narrow, slit-like spaces inside the lobes, and the almost solid stem are characteristic of this mushroom. The head is dark, either brownish or purple-brown, a different colour from that typical of morels.